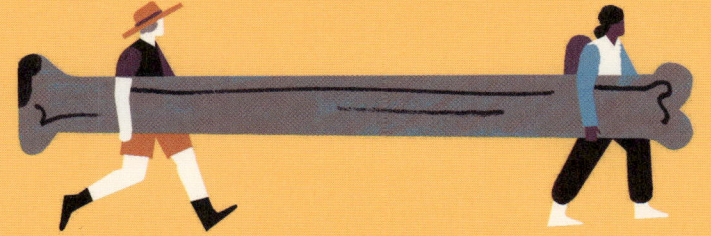

DINOSAUR DYNASTY

For Ffi and our little pack – Norman and Coen

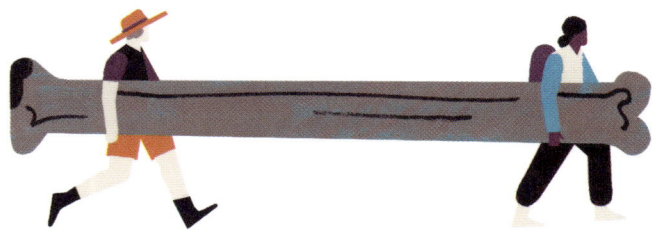

BIG PICTURE PRESS

First published in the UK in 2025 by Big Picture Press,
an imprint of Bonnier Books UK
5th Floor, HYLO, 105 Bunhill Row,
London, EC1Y 8LZ
The authorised representative in the EEA is
Bonnier Books UK (Ireland) Limited.
Registered office address:
Floor 3, Block 3, Miesian Plaza,
Dublin 2, D02 Y754, Ireland
compliance@bonnierbooks.ie
www.bonnierbooks.co.uk

Copyright © 2025 by Jack Tite

1 3 5 7 9 10 8 6 4 2

All rights reserved

ISBN 978-1-80078-988-3

This book was typeset in Graham and Quicksand
Illustrations were created digitally using shapes, colour and texture

Written by Jack Tite
Consulted by Dougal Dixon and Kimberley Davis
Designed by Jack Tite & Sarah Crookes
Edited by Josephine Southon

Printed in China

DINOSAUR DYNASTY

Written & illustrated by Jack Tite

BPP

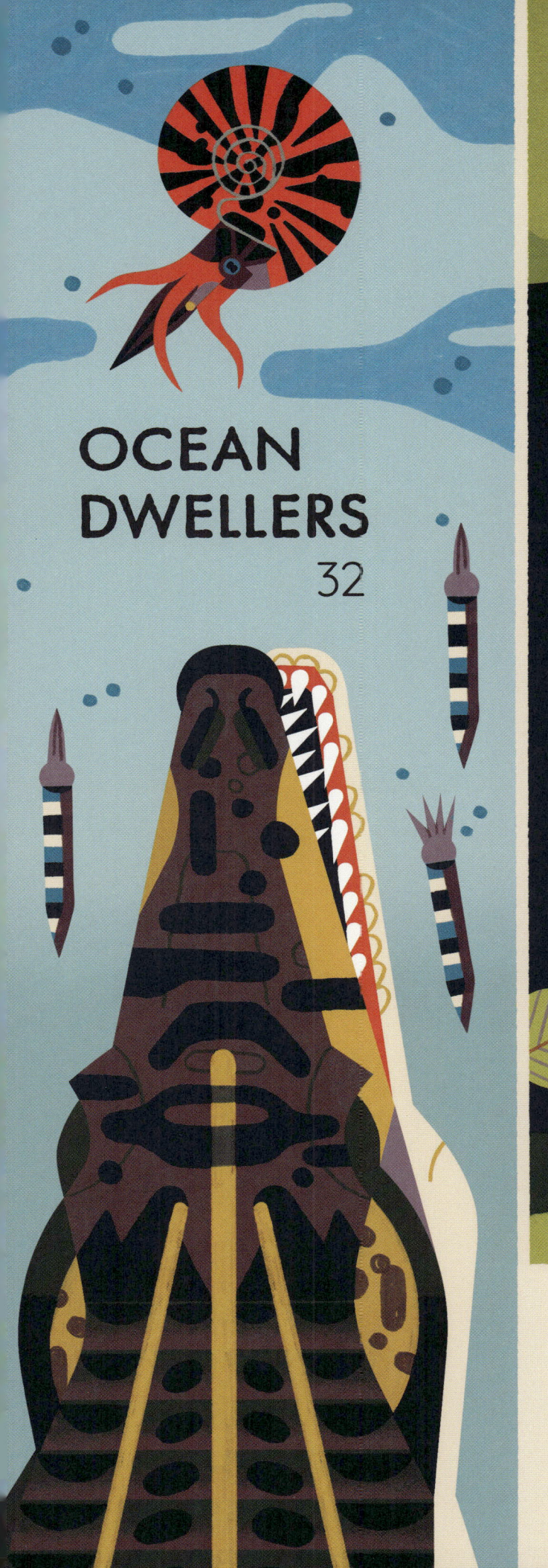

OCEAN DWELLERS 32

CRETACEOUS 38

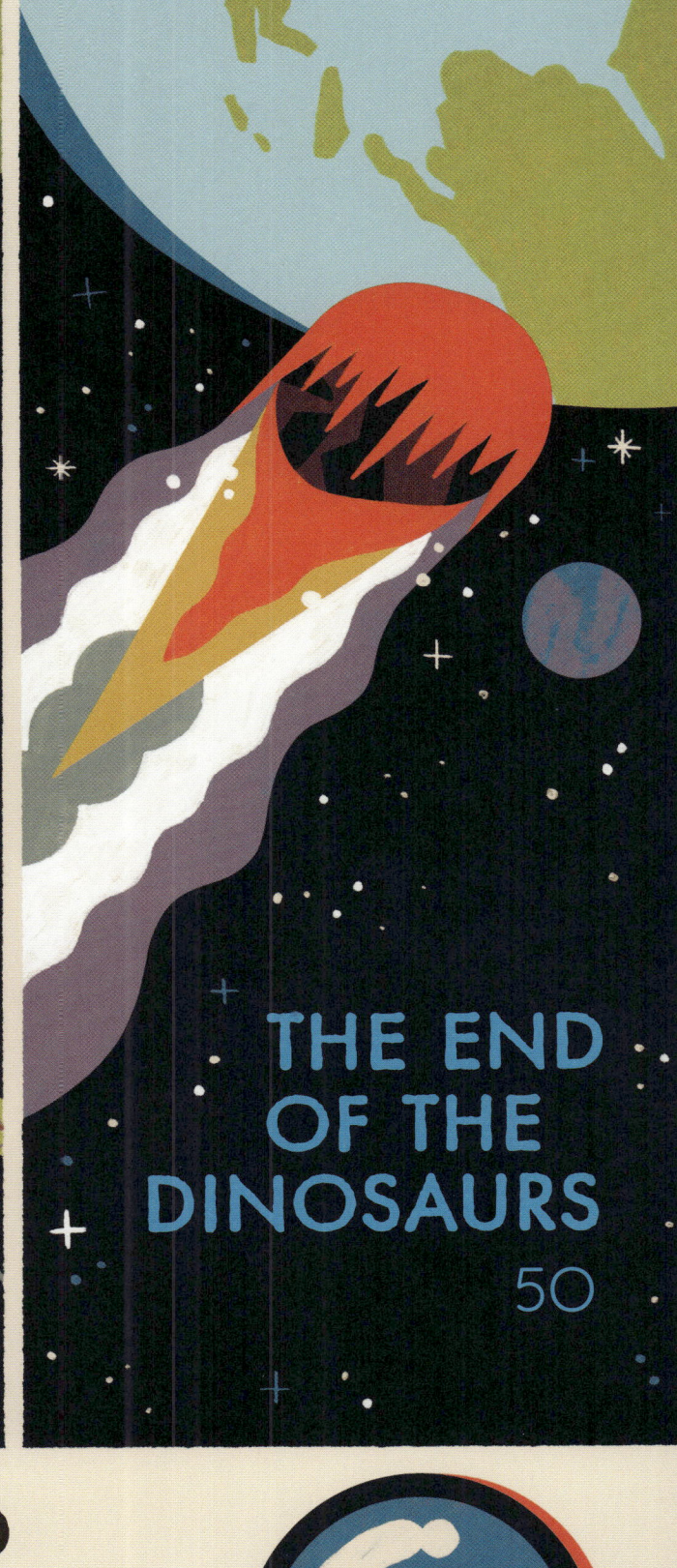

THE END OF THE DINOSAURS 50

GLOSSARY AND PRONUNCIATIONS 60

THE AGE OF REPTILES

Hundreds of millions of years ago **dinosaurs** ruled the planet. This period, from around 252 million to 66 million years ago, is known as the **Mesozoic Era**. Soaring **reptiles** armed with sharp beaks and claws streamed through the skies, the oceans were teeming with snappy prehistoric **predators** and the land belonged to a vast array of mighty dinosaurs.

Hitting the big time
The Mesozoic Era lasted 186 million years, more than 600 times longer than modern humans have been on Earth, a mere 300,000 years. This is split up into three periods: **Triassic**, **Jurassic** and **Cretaceous**. During this epic chunk of time, some animals died out and others changed to survive. This is called **evolution**.

Triassic Earth Jurassic Earth Cretaceous Earth

The moving world
In the beginning of the Mesozoic Era, all land on earth was joined together to form a giant supercontinent known as **Pangaea**. The land – and all the creatures living on it – drifted apart during the next 186 million years to look more like the Earth as it is today. Back then, our planet was warmer, with vast desert plains, flourishing forests and everchanging habitats due to the moving land.

A mixed bag
Scientists now know of over 1,000 different **genera** of dinosaur, with around 50 new genera discovered each year. There were speedy sprinters, gentle giants, armoured titans and formidable predators. They ranged from tiny reptiles that could sit on your finger to aeroplane-sized giants that proudly stand in today's museums. Our ancient ancestors lived alongside the dinosaurs, but they were small, simple, rodent-like mammals that survived in the shadows of the mighty reptiles.

TRIASSIC

From 252 to 201 million years ago, Pangaea was mostly dry and warm, with windy monsoons sweeping across the planet from time to time. Towering forests, arid desert plains and vast fern prairies provided homes for early amphibians and reptiles.

EUROPE & ASIA

NORTH AMERICA

AFRICA

SOUTH AMERICA

AUSTRALIA

ANTARCTICA

WHAT CAME BEFORE

The Triassic Period began with the largest ever mass **extinction** around 252 million years ago, when at least 90% of all life on Earth was wiped out. Entire groups of reptiles, early mammals, insects and aquatic life disappeared, but some creatures survived. These included the ancestors of crocodiles and dinosaurs, known as archosaurs and early mammals called synapsids.

Insects arrived over 230 million years before the Triassic Period began.

Roughly the size of a wolf, *Cynognathus* was a meat-eating synapsid with sharp teeth and a large, strong jaw.

Lystrosaurus was a synapsid the size of a large dog. It used its turtle-like beak to snap up plants and dual tusks to dig burrows or root out food beneath the earth.

Aetosaurs resembled today's crocodiles, but with tough, bony armour and small heads. Some had shoulder spikes, some ate plants and a few ate meat.

Ruling the skies

Towards the end of the Triassic, a group of archosaurs took to the skies, tens of millions of years before birds. These are now known as pterosaurs. The outer structure of a pterosaur wing was formed by one long finger, with a thin layer of skin stretching from the fingertip to the back legs. Their soft-shelled eggs (similar to today's snake eggs) have been discovered **fossilised** in ancient nesting grounds where colonies of pterosaurs once nurtured their young.

Early pterosaurs, such as the gull-sized *Caelestiventus*, likely ate insects and fish or scavenged dead animals in what is now western North America. Smaller long-tailed pterosaurs include *Peteinosaurus* and *Eudimorphodon*, which soared across present-day Europe.

Making waves

At some point in their prehistoric past, certain land reptiles slowly evolved to master life in the oceans instead. Ichthyosaurs looked like a cross between a crocodile and dolphin, with flippers, a wide tail and a long, thin snout with plenty of teeth for eating fish, squid and shellfish. The smallest ichthyosaurs were about the length of an otter, whilst the largest matched the size of a whale.

Swimming with sharks

Sharks lived long before the Triassic Period and thrived during the Mesozoic Era. Some were armed with sharp teeth for grasping fish or squid, whilst others had rounded back teeth for crushing shelled creatures.

Plesiosaurs were another group of aquatic reptiles. They used their long necks and powerful bite to snap up **prey** from the sea floor.

Tanystropheus was a fish-eating archosaur with an extremely long, winding neck and overlapping jagged teeth.

Snappy, pig-sized, plant-eating rhynchosaurs were a tasty treat for many predators of the time.

HUMBLE BEGINNINGS

The earliest dinosaurs are thought to have arrived around 240 million years ago, in the Middle Triassic Period. Instead of toppling the ruling archosaurs, they lived alongside them, probably falling prey to the larger predators of the time. We know of a few early dinosaur **species** that the later giants evolved from.

Brain size of a Tyrannosaurus rex

Dinosaurs had small brains compared to their body size. Scientists believe they behaved like today's crocodiles and lizards.

Eoraptor (E. lunesis)

This is *Eoraptor*, discovered in Argentina, South America. It most likely used its sharp, leaf-shaped front teeth and curved, saw-like back teeth to eat plants, small lizards and insects. *Eoraptor* stood on two legs and was about the size of a small dog. It was built for speed, with a long neck, skinny frame and powerful back legs. It had five-fingered hands, though only three of these fingers had claws; the other two were basically useless. Dinosaur claws were made from a type of **keratin**, the same material in our hair, nails and rhinoceros' horns.

Eoraptor compared to a human and small dog.

Eoraptor diet

We don't know what *Eoraptor*'s skin looked like, but it was probably scaly like many other dinosaurs. Later dinosaurs evolved dangerous spiked spines, bony armour plates called **osteoderms**, and even funky feathers.

Like most lizards, dinosaurs laid eggs to reproduce. The eggs varied in shell colour, size and thickness depending on the dinosaur. Even the largest dinosaurs laid eggs that were only about the size of an ostrich egg.

The hips don't lie

Dinosaurs branched out and evolved rapidly from early species such as *Eoraptor*. Later dinosaurs came in all shapes and sizes, with all kinds of diet, teeth, feet, defence mechanisms and an arsenal of weapons. They are divided into two groups based on their pelvic bones, which differed from one another. **Saurischians** included four-legged, long-necked dinosaurs and deadly **carnivores**, such as *Tyrannosaurus rex*. The **ornithischians** included the spiky *Stegosaurus* and *Triceratops*.

Eoraptor had hollow bones, making them incredibly light yet strong, a trait that many dinosaurs benefited from. With legs positioned beneath their bodies, dinosaurs could grow taller and move quicker than today's reptiles, whose shorter legs extend out to the sides.

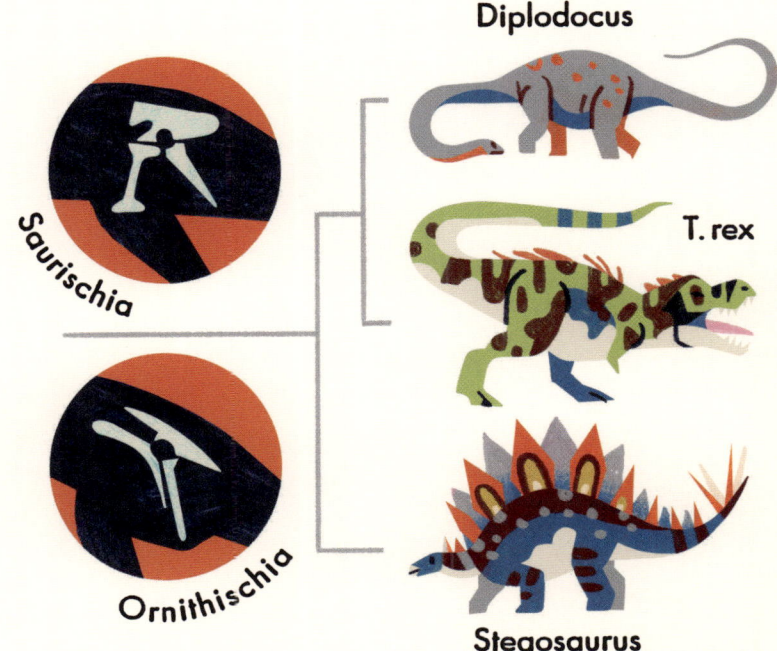

HISTORY REPEATING ITSELF

Growing up to 8 metres long, *Plateosaurus* towered over its peers at the end of the Triassic Period, in what is now Europe. This elephant-sized plant-eater may have moved in **herds** to protect its young from hungry meat-eaters such as the 5-metre-long *Liliensternus*.

American remains

In Argentina, the meat-eating *Herrerasaurus* was a swift and deadly hunter, standing as tall as a person and 3 metres long. Some dinosaurs were smaller but no less important. The slightly smaller *Coelophysis* hunted insects and small, ground-dwelling creatures across modern North America and Africa. It was a lightweight little hunter about half the size of *Herrerasaurus*. In New Mexico, USA, a site called Ghost Ranch has yielded a bountiful bonanza of nearly 1,000 *Coelophysis* bones.

Sauropods and theropods

All these dinosaurs are early saurischians. *Plateosaurus* belonged to a group that later contained the **sauropods**, known for their long necks, thick legs and large bodies. *Liliensternus*, *Herrerasaurus* and *Coelophysis* are **theropods**, a meat-eating group with mostly hollow bones, claws and curved teeth.

A golden opportunity

For reasons not fully known, the Triassic Period ended around 200 million years ago with another huge extinction event. Whilst this wiped out around 75% of animals on land and in the sea, it helped the dinosaurs. Less competition allowed them to muscle their way into new places, take a hold and thrive. This marks the end of the Triassic Period and the beginning of a new era in which the dinosaurs ruled.

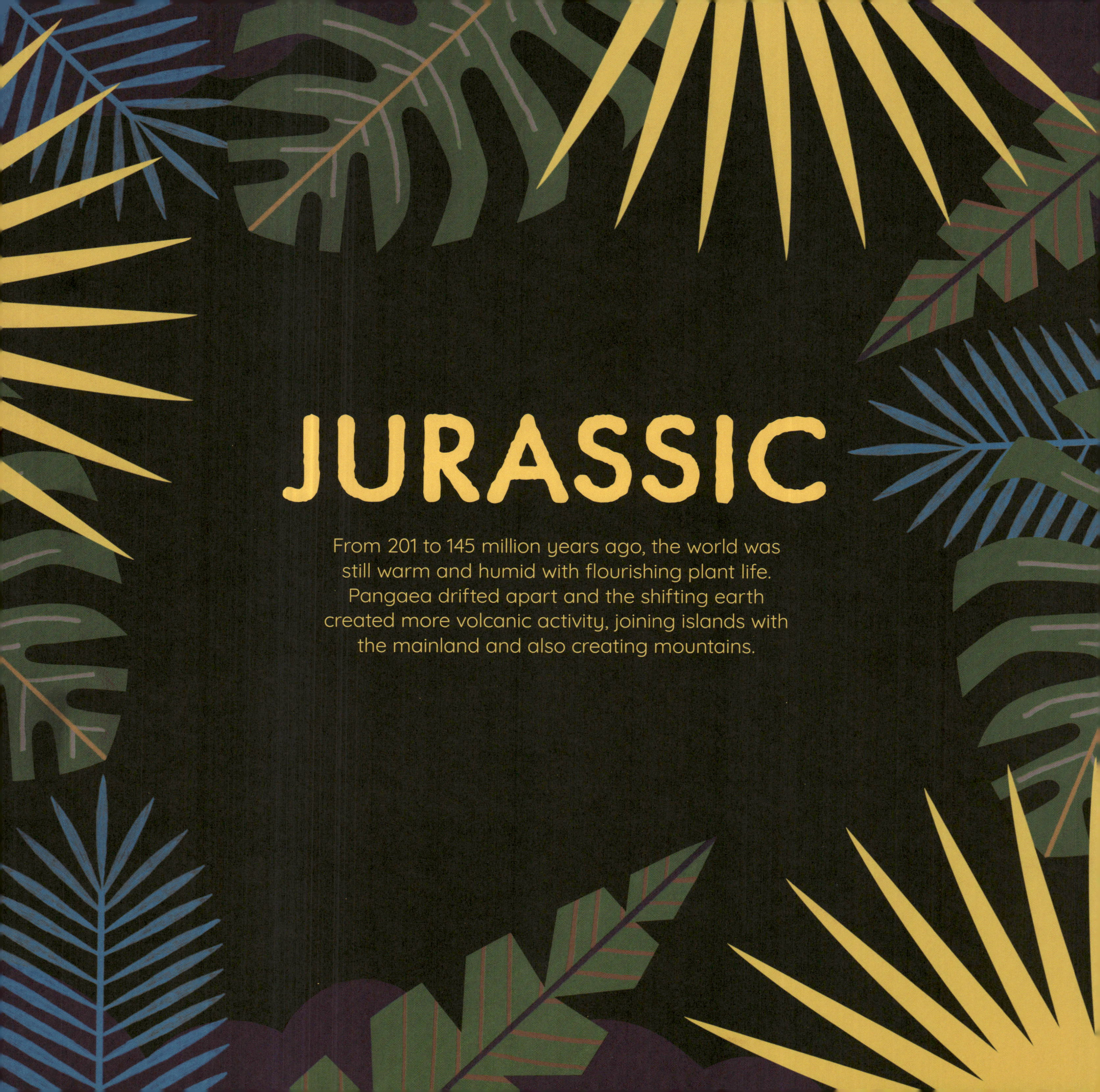

JURASSIC

From 201 to 145 million years ago, the world was still warm and humid with flourishing plant life. Pangaea drifted apart and the shifting earth created more volcanic activity, joining islands with the mainland and also creating mountains.

NORTH AMERICA

EUROPE & ASIA

AFRICA

SOUTH AMERICA

AUSTRALIA

DAWN OF THE DINOSAURS

Now it was the dinosaurs' time to shine. The Jurassic Period began 201 million years ago, a time when the world was transforming and Pangaea was beginning to drift apart. It was warm, with no frozen poles like Antarctica today, and plants flourished to provide plentiful food for **herbivores**. Insects scuttled all across the humid planet, including beetles, dragonflies, spiders and the first wasps and ants. Various different species of pterosaur sprung from the Jurassic Period, including pigeon-sized, furry *Anurognathus* and its cousins. These bug-eyed **ambushers** popped up in the Middle Jurassic Period across Europe and Asia, probably hunting at night like bats.

Anurognathus

Diverse flyers

The earliest pterosaurs were the long-tailed, narrow-winged, short-necked rhamphorhynchoids, and then in the Late Jurassic they branched off into the short-tailed, broad-winged, long-necked pterodactyloids. The seagull-sized *Pterodactylus* is one of the best known from this group, soaring in the skies of present-day Europe and Africa 150 million years ago in search of small animals to eat. The largest rhamphorhynchoid, *Dearc*, soared above what is now Scotland, snapping up fish with its toothy beak. With a wingspan of 3.8 metres, it was just larger than the wandering albatross, which has the largest wingspan of any bird alive today. The much smaller *Dimorphodon*, with a hefty yet lightweight skull, fed on small creatures and possibly fish during the Early Jurassic years.

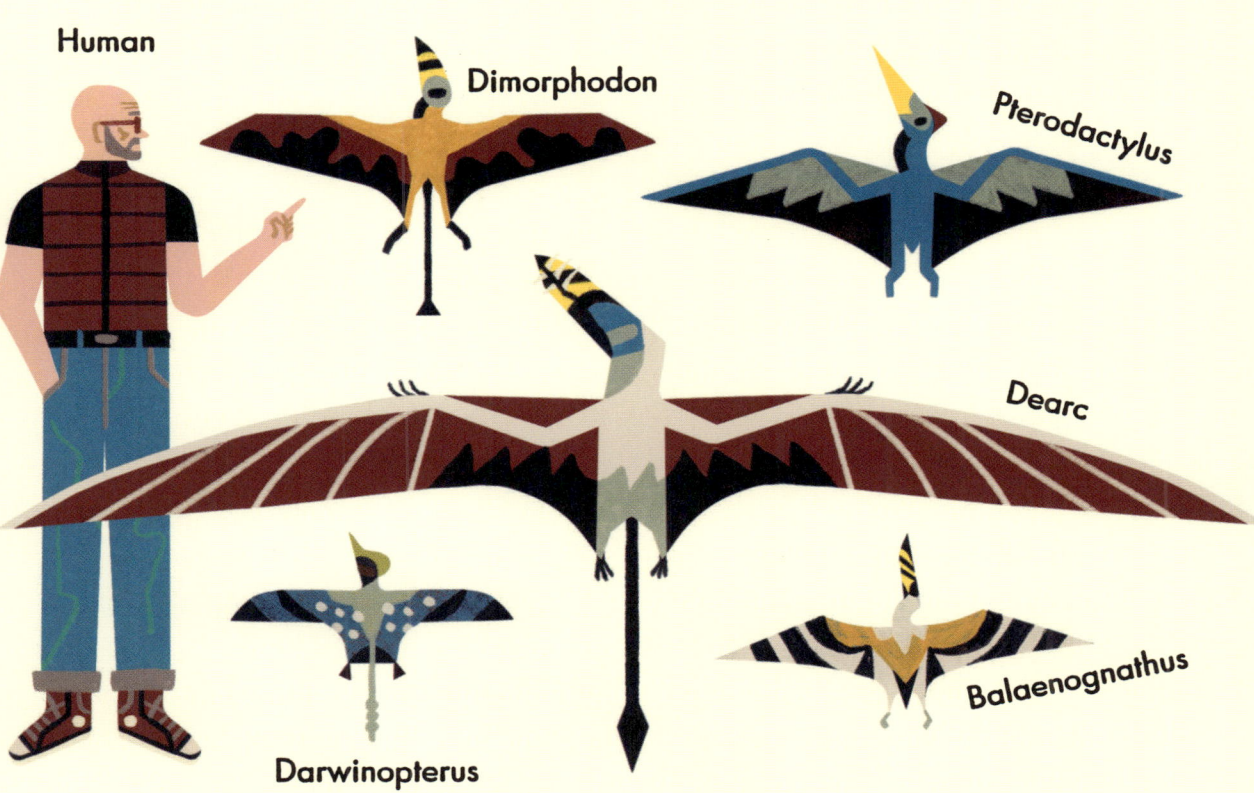

Born ready

Just as many male and female birds look different to one another today, so did many pterosaurs. In what is now China, crow-sized *Darwinopterus* males had a larger **crest** than the females, which served to attract mates or scare off rivals. The females laid many eggs, each one weighing as little as a pencil. Unlike birds, the hatchlings were able to take to the skies straight away.

The filter-feeder

An odd-looking pterosaur called *Pterodaustro* was just larger than a duck and waded in South American shallow shorelines during the Early Cretaceous. Like today's whales and flamingos, they combed through the water with hundreds of long, thin teeth in their beak-like mouths to catch tiny creatures.

BORN TO BATTLE

Jurassic predators and prey were constantly evolving to outdo one another. Standing up to 2 metres tall and 7 metres long, *Dilophosaurus* is widely considered as the top predator of the Early Jurassic Period. Its name, meaning 'two crested lizard', comes from the pair of crests sitting atop its skull. These crests were probably colourful and used to attract mates or recognise each other. *Dilophosaurus* was swift and agile, prowling North America with long sharp claws and serrated teeth to take down prey. Some scientists think that it may have scavenged, some argue it ate fish. Bite marks on the bones of a *Sarahsaurus* suggest it was actively hunted by the *Dilophosaurus*.

Hands down
Sarahsaurus was about the size of a car, walking on two legs with sturdy hands for grabbing lush greens, just like *Plateosaurus* of the Triassic Period. Relatives across the globe began to walk on all fours and grow even larger during the Early Jurassic, such as the bus-sized *Barapasaurus*, found in what is now India.

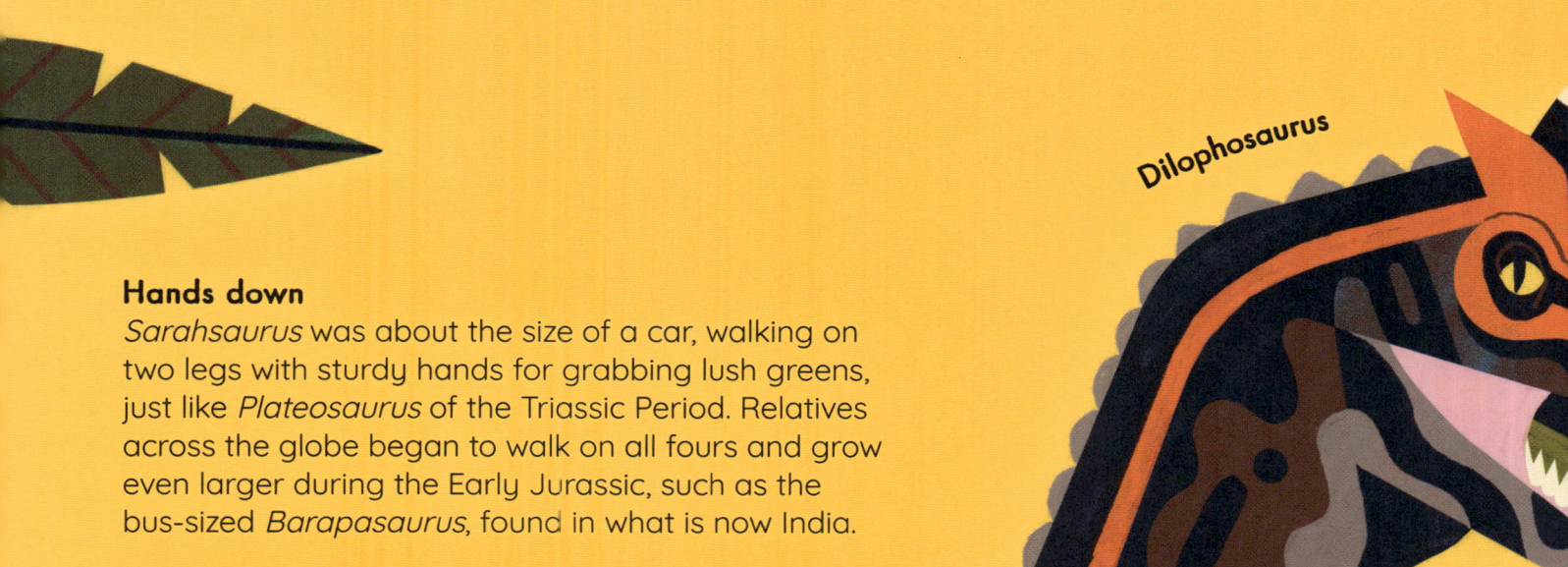
Dilophosaurus

Shield bearer

Scutellosaurus lived alongside these two foes. Measuring up to 2 metres in length, it walked on two legs with a slim, small body shielded by hundreds of osteoderms. It is the earliest known thyreophoran (meaning 'shield bearer'), a group of plant-eating armoured dinosaurs. A close relative called *Scelidosaurus* lived at the same time and measured up to 4 metres long, about the same length as a lion. Later, bulky creatures from this group would grow even bigger and don more elaborate bony armour or deadly spikes.

LAND OF GIANTS

All across the planet, sauropods evolved features that turned them into towering giants. Walking on all fours with pillar-like legs was the new sauropod norm, with long, spindly necks to access food out of reach to others. Outstretched necks meant that winding tails were also needed to balance their bodies.

Featherweight champions
Sauropods had spaces between and around their bones filled with air sacs to store oxygen when breathing – a feature that modern birds also have. This made their skeletons featherweight yet sturdy enough to carry their hulking body.

Safety in numbers
Scientists think that many sauropods lived in herds to protect their young. This, along with being big, was probably enough to scare away any daring hunters. If not, a long tail could have been a deadly whipping weapon when needed.

Sauropod genera
From their dog- and giraffe-sized ancestors, different sauropod genera swept across the ever-changing planet during the Jurassic Period.

Over 300 bones from six different *Barapasaurus* individuals have been found in India dating from the Early Jurassic Period – around 190 million years ago. It was an early sauropod that grew up to 14 metres long.

Downsizing
Some animals shrink over time due to a smaller environment, such as being trapped on an island. This is called **insular dwarfism**. In what is now Germany, the island-dwelling *Europasaurus* was among the smallest of the sauropods, with its body about the same size as a cow.

SMALL AND MIGHTY

Some dinosaurs that didn't grow large found other innovative ways to survive. Somewhat of a trendsetter, *Archaeopteryx* was among the first handful of feathered dinosaurs that took to the skies. This theropod had a long tail and cone-shaped teeth to munch on lizards, small mammals and possibly insects. Around 150 million years ago, it flew above a cluster of tropical islands in what is now Germany.

Flight feather structure

Archaeopteryx

Compsognathus

Life-size *Compsognathus* head

Pint-sized powerhouse
This tiny speedster is *Compsognathus*, a chicken-sized predator that shared the islands with *Archaeopteryx*. Sharp teeth and claws were effective weapons against any small prey it chased down.

Working up an appetite
The largest sauropods are estimated to have needed 450 kilograms of food a day – about the weight of 1500 lettuces. With no back teeth or cheeks, they couldn't chew food, so they ate everything whole. Like many birds and crocodiles, sauropods swallowed stones, which sat in their guts to slowly grind and break apart food. Slicing at tough plants all day eventually wore sauropod teeth down, so they needed replacing to keep up with their hungry appetites. *Diplodocus* could replace a tooth once every 35 days.

Diplodocus teeth (far left)
Camarasaurus teeth (closest left)

Life-size sauropod egg

Heterodontosaurus

Armed to the teeth
Heterodontosaurus was truly unique because it had three different types of teeth – small, blade-like front teeth, sharp tusks and square back teeth for grinding – which it used to feast on plants and the occasional insect. It stood about the size of a fox and prowled the deserts looking for lush lakes and food 200 million years ago, in what is now Southern Africa.

Incisivosaurus

Long in the tooth
The rather odd *Incisivosaurus* had two large front teeth and rows of peg-shaped back teeth, suggesting a plant-based diet. It likely sported feathers and lived around 125 million years ago in what is now China.

Triceratops, a later ceratopsian

Yinlong

What a bonehead
Yinlong (meaning 'hidden dragon') is another very important dinosaur that lived around 159–154 million years ago, in what is now China. It's the earliest known ceratopsian (an important group later on), with a bony skull and plant-snipping, beak-like mouth. It had four-fingered hands, stood on two legs and was roughly the size of a large turkey.

APEX PREDATORS

When plant-eaters grow large and learn new tricks, so do predators. The Jurassic Period saw the rise of terrifying meat-eaters. They stood on two powerful legs with a long tail for balance and hefty jaws lined with deadly teeth.

Torvosaurus

Ceratosaurus

A deadly double whammy
The truck-sized, 10-metre-long *Torvosaurus* hunted across what is now North America and parts of Europe. A thick body, sharp teeth and huge claws helped it take down large herbivores 150 million years ago. About half its size but with pointier features, *Ceratosaurus* prowled the same hunting grounds. It bore a horn above each eye and its nose, and had bony ridges along its back. It attacked its prey with long fangs, sharp claws and strong arms.

Big Al

Completing this deadly trio of American carnivores is *Allosaurus*. This horned dinosaur lived in what is now North America, as well as Africa and parts of Europe. It was among the largest predators of the time, with one skeleton nicknamed 'Big Al' by scientists. The head alone was about the size of a coffee table. It had curved teeth that it shed and regrew, and it sported pointy horns above each eye. Strong arms and curved claws were used for slashing. All this firepower and a length of up to 12 metres helped *Allosaurus* fight its way to the top of the food chain.

Two sides of the same coin

In what is now China, a very similar story took place during the Jurassic Period. *Yangchuanosaurus* was a similar size to *Allosaurus*, and spent its days hunting enormous sauropods and relatives of *Stegosaurus*. These had slightly different armoured plates and builds, such as the smaller *Chungkingosaurus* and its cow-sized cousin, *Tuojiangosaurus*.

Not alone

There were a few different spiky or armour-plated dinosaurs during the Jurassic. Alongside these predatory enemies lived the herbivorous *Gargoyleosaurus* – standing roughly the height of a large dog and 4 metres long. Reinforced armoured plating protected its back whilst it scoured the floor for low-lying plants.

The ultimate showdown

The bus-sized *Stegosaurus* is another superstar from the Jurassic Period, clipping away at tasty plants. Its tall plates were probably used to intimidate help identify other *Stegosaurus*, or even to store heat like radiators. A fully *Stegosaurus* measured up to 9 metres long and defended itself with a swo adorned with four spikes, making it a true challenge for even the mighty A

OCEAN DWELLERS

Through the entire Mesozoic Era, the giant ocean called **Panthalassa** slowly separated to resemble the oceans today. By the Jurassic Period, a variety of marine creatures thrived. Massive coral reefs lined the ocean floor built by a variety of **molluscs** called rudist clams, which are related to today's mussels.

Polar predator
Cryolophosaurus sported a crest atop its head and stood as tall as a large polar bear on its hind legs at 2.5 metres. Antarctica was its home, but the continent wasn't the frozen plain it is today. Back in the Early Jurassic Period 170 million years ago, Antarctica was a lush, warm woodland teeming with sauropods and small mammals for this carnivore to hunt down.

ABUNDANT SEAS

At the dawn of the Mesozoic Era, one enormous ocean called Panthalassa surrounded Pangaea. As the land shifted apart and new oceans formed, ichthyosaurs and plesiosaurs swarmed the seas, but other more familiar creatures also lurked beneath the waves.

Life at sea
During the Jurassic and Cretaceous Periods, oceans were teeming with crabs, lobsters and rays. Creatures called ammonites also thrived – they were similar to squids but they had a variety of distinctive shell shapes and sizes. Some shells were the size of a coin while others had huge, broad shells wider than a table. They fed on small, slow critters and ocean plants, but fell prey to marine reptiles and sharks.

Famished fish

A fierce predatory fish called *Xiphactinus* dwarfed some sharks, growing up to 5 metres long. It gobbled up anything that would fit in its fang-lined jaws, such as other large fish, sharks and squid.

Dinosaurs for dinner

Some sharks snapped up any prey they could get their jaws on. *Cretoxyrhina* and *Squalicorax* of the Creteaceous Period resembled today's great white sharks, and they ate turtles, other sharks, plesiosaurs, large fish, seabirds and some smaller dinosaurs.

Titanic turtles

At the end of the Cretaceous Period lived the gigantic turtle *Archelon*. The same size as a small car, it trawled along the seafloor in search of jellyfish, squid, ammonites, fish and plants.

BEASTS BELOW

Reptiles evolved into unstoppable predators during their time at sea. **Mosasaurs** had snake-like bodies and propelled themselves through the water with their tail and wide, paddle-like flippers. The smallest were a meagre 1 metre long, but the largest grew up to 17 metres in length – the size of a humpback whale today. These giants hunted for ammonites, fish, plesiosaurs and turtles.

Changing shape
Some plesiosaurs branched off, ditching their long necks and supersizing their heads to transform into well-built beasts called pliosaurs. A giant of this group, measuring up to 10 metres long, is known as *Kronosaurus*. It glided through the Cretaceous oceans by moving all four flippers at once, pursuing turtles, plesiosaurs or ammonites and snapping them up in its powerful jaws. Embedded in these jaws were dozens of long, deadly teeth, the largest of which measured up to 30 centimetres long!

Life-size *Kronosaurus* tooth

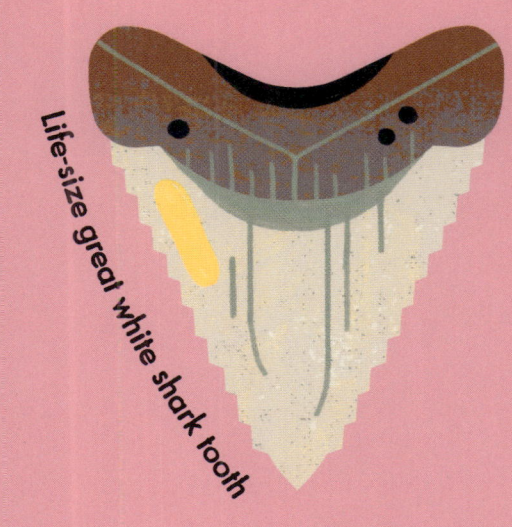

Life-size great white shark tooth

A big bite
Mosasaurus's car-sized skull housed 40–50 cone-shaped teeth, each growing as long as 7–10 centimetres – a similar length to a great white shark's tooth. Mosasaurus had jaws that could open up extremely wide to swallow large prey whole, similar to some snakes today. Also like snakes, a second set of smaller teeth lined their throat to stop prey from escaping.

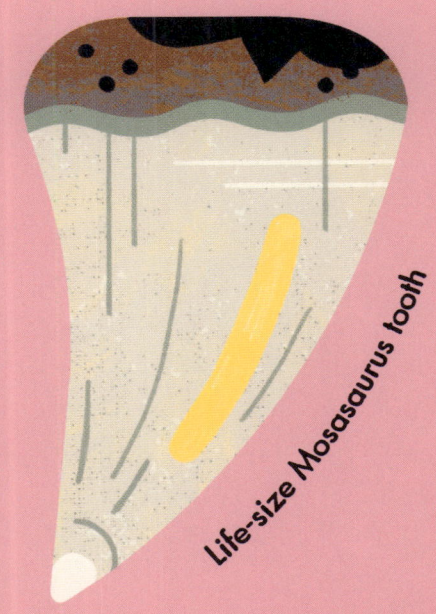

Life-size *Mosasaurus* tooth

CRETACEOUS

The Cretaceous Period began some 145 million years ago as the land shifted closer to how it is today. This was a time of continued growth and change for dinosaurs, giving rise to the largest, most deadly and most bizarre beasts yet. The earliest flowering plants came into their own, adding a splash of colour to the wave of green plant life.

NORTH AMERICA

EUROPE & ASIA

AFRICA

SOUTH AMERICA

ANTARCTICA

AUSTRALIA

ARMOURED AND DEADLY

When the Cretacous Period began, land masses continued to move, leading to new oceans being formed. The first flowering plants sprouted during this time, bringing colour and diversity to the pre-existing plant life and prompting another explosion in insect life. Dinosaurs continued to evolve in ingenious ways to help them thrive.

Built like a tank
One of the largest and most well-loved dinosaurs today is *Ankylosaurus*. A battle-ready, 10-metre-long fortress, it outweighed even the African elephant. Its body was covered in a mosaic of bony, spike-studded armour, with a tough helmet fused into its skull and a deadly clubbed tail. It roamed around what is now North America 68–66 million years ago, using its keen sense of smell and sharp beak to pluck at foliage on the forest floor. Wielding a bone-breaking wrecking ball of a tail, this tank may have fought other rival **ankylosaurs** or fended off daring predators, such as *T. rex*.

DRESSED TO IMPRESS

Among the vast array of herbivorous herds roaming the planet were many dinosaurs decorated with **frills**, horns or other features to attract mates, fend off hungry predators or bellow out booming calls.

Centrosaurus

Chasmosaurus

Horns and all
Hailing from the same family as the rather simple Jurassic *Yinlong*, ceratopsians mostly spread across America and Asia for the entire Cretaceous Period. They were bulky reptiles that stomped on all fours, and ranged from the size of a dog to the size of a large van. Most had plant-picking beaks and elaborate, bony neck frills that varied widely.

Styracosaurus

Over 60 different ceratopsian genera are known, each equipped with outlandish headgear; from the sheep-sized *Protoceratops*, living in what is now Asia, 74–70 million years ago, to the highly decorated North American *Kosmoceratops* that lived 76 million years ago. It had 10 overhanging hooked spikes, two curved brow spikes and a blunt nose horn.

Kosmoceratops · Protoceratops · Einiosaurus · Nasutoceratops · Achelousaurus · Sinoceratops

Assorted armour
Today, scientists know of well over a dozen different types of ankylosaurs. Some had long spikes sprouting out from their necks and shoulders, such as the hippopotamus-sized *Gastonia* and slightly larger 6-metre-long *Sauropelta*, both of which lived in what is now North America. Others were smaller with lighter, simpler armour, such as the sheep-sized *Minmi* that lived around 120–112 million years ago in present-day Australia.

Ankylosaurus

Sauropelta

Gastonia

Minmi

Frilled and spiky
Styracosaurus had very impressive headgear, with as many as six long spikes, horned cheeks, a beak-like mouth and a thick nose horn that grew up to 60 centimetres long. It was built like a rhinoceros but larger, at up to 5.5 metres long. *Styracosaurus* probably lived in groups that roamed present-day North America, 76–70 million years ago. It also lived alongside other horned hunks with slightly different frills, *Centrosaurus* and *Chasmosaurus*.

Big Al

Completing this deadly trio of American carnivores is *Allosaurus*. This horned dinosaur lived in what is now North America, as well as Africa and parts of Europe. It was among the largest predators of the time, with one skeleton nicknamed 'Big Al' by scientists. The head alone was about the size of a coffee table. It had curved teeth that it shed and regrew, and it sported pointy horns above each eye. Strong arms and curved claws were used for slashing. All this firepower and a length of up to 12 metres helped *Allosaurus* fight its way to the top of the food chain.

Not alone

There were a few different spiky or armour-plated dinosaurs during the Jurassic. Alongside these predatory enemies lived the herbivorous *Gargoyleosaurus* – standing roughly the height of a large dog and 4 metres long. Reinforced armoured plating protected its back whilst it scoured the floor for low-lying plants.

The ultimate showdown

The bus-sized *Stegosaurus* is another superstar from the Jurassic Period, clipping away at tasty plants. Its tall plates were probably used to intimidate help identify other *Stegosaurus*, or even to store heat like radiators. A fully *Stegosaurus* measured up to 9 metres long and defended itself with a sw adorned with four spikes, making it a true challenge for even the mighty A

Two sides of the same coin

In what is now China, a very similar story took place during the Jurassic Period. *Yangchuanosaurus* was a similar size to *Allosaurus*, and spent its days hunting enormous sauropods and relatives of *Stegosaurus*. These had slightly different armoured plates and builds, such as the smaller *Chungkingosaurus* and its cow-sized cousin, *Tuojiangosaurus*.

OCEAN DWELLERS

Through the entire Mesozoic Era, the giant ocean called **Panthalassa** slowly separated to resemble the oceans today. By the Jurassic Period, a variety of marine creatures thrived. Massive coral reefs lined the ocean floor built by a variety of **molluscs** called rudist clams, which are related to today's mussels.

Polar predator
Cryolophosaurus sported a crest atop its head and stood as tall as a large polar bear on its hind legs at 2.5 metres. Antarctica was its home, but the continent wasn't the frozen plain it is today. Back in the Early Jurassic Period 170 million years ago, Antarctica was a lush, warm woodland teeming with sauropods and small mammals for this carnivore to hunt down.

The meat bull boogie
The fearsome, 8-metre-long *Carnotaurus* (meaning 'meat bull') hunted big game in what is now Argentina around 71–69 million years ago. It is easily recognised by two horns, one above each eye. Its two barely noticeable arms were highly flexible, possibly helping win over females with a whirly and wacky dinosaur dance.

Abelisaurids
Abelisaurids are another group of meat-eating theropods that lived in what is now Africa, India, South America and possibly Europe. They typically had surprisingly small arms, short-faced skulls, small sharp teeth, muscular necks and athletic builds for ambushing prey.

Rajasaurus

India's king lizard
In today's India, 72–66 million years ago, *Rajasaurus* (meaning 'king lizard') was another large predatory dinosaur, measuring up to 6.6 metres long. It sported a single small horn atop its head that may have been used for bashing rivals or for showing off to females.

THE BIRDS AND THE BEASTS

Feathers really caught on during the Cretaceous Period. In Asia during the early parts of the era, the ceratopsian *Psittacosaurus* flaunted an odd bristly tail. Many other creatures would go on to don fully fledged feathers and fly.

Psittacosaurus

Gallimimus

Gallimimus lived in present-day Mongolia, Asia, 70 million years ago. It stood over twice as tall as an adult human, with a long, duck-billed snout for eating small creatures, plants and the occasional egg. Known for their ostrich-like bones, *Gallimimus* belonged to a group of speedsters called ornithomimosaurs, or bird-mimics.

Gigantoraptor

Oviraptor

Avimimus

A feathered family of theropods called oviraptorosaurs adopted parrot-like beaks and spread across present-day Asia and North America 130–66 million years ago. They ranged from the small, dog-sized *Oviraptor* and *Avimimus* to the giraffe-sized *Gigantoraptor*, which weighed as much as a white rhino.

Dromaeosaurs flourished all over the globe during the Cretaceous Period. They boasted sharp claws and teeth, a strong bite, long tails and most likely had feathers. These raptors hunted in packs, pinning down prey with a huge, hooked toe claw. This group included the turkey-sized *Velociraptor*, found in present-day Asia, and the much larger *Deinonychus*, found in the United States of America. In China, the pigeon-sized *Microraptor* shared the skies with pterosaurs and other feathered flyers.

The peculiarly fuzzy *Therizinosaurus* stood taller than a giraffe at up to 7 metres tall. It had the longest claws of any known animal, measuring a whopping 1 metre long, which it used to grasp plants in Mongolia at the end of the Cretaceous Period.

Microraptor

Deinonychus

Velociraptor

Therizinosaurus

Amargasaurus

Concavenator

Sails, which were large, flat extensions of a dinosaur's spine, remained a common feature for some species. Around 130 million years ago, the deadly *Concavenator* had a small sail, hunting around what is now Spain. The strange 10-metre-long sauropod *Amargasaurus* roamed the arid plains in today's Argentina.

THE BIG SHOTS

On land the most enormous plant-eating reptiles towered over the prehistoric world like walking skyscrapers, with epic predators emerging to hunt them. In the air, the skies would have been blackened by some of the largest flying animals to have ever lived streaming like aeroplanes up above.

Is it a bird, is it a plane?
Quetzalcoatlus had wings like a fighter jet and stood at a similar height to a giraffe. It waded in the rivers of what is now Texas, USA, around 67 million years ago, snatching fish with its long toothless **bill** and launching into the sky to soar on the winds above. At the same time on tropical islands in today's Romania, Europe, the similarly sized *Hatzegopteryx* probably feasted on dwarf sauropods along with an array of downsized dinosaurs.

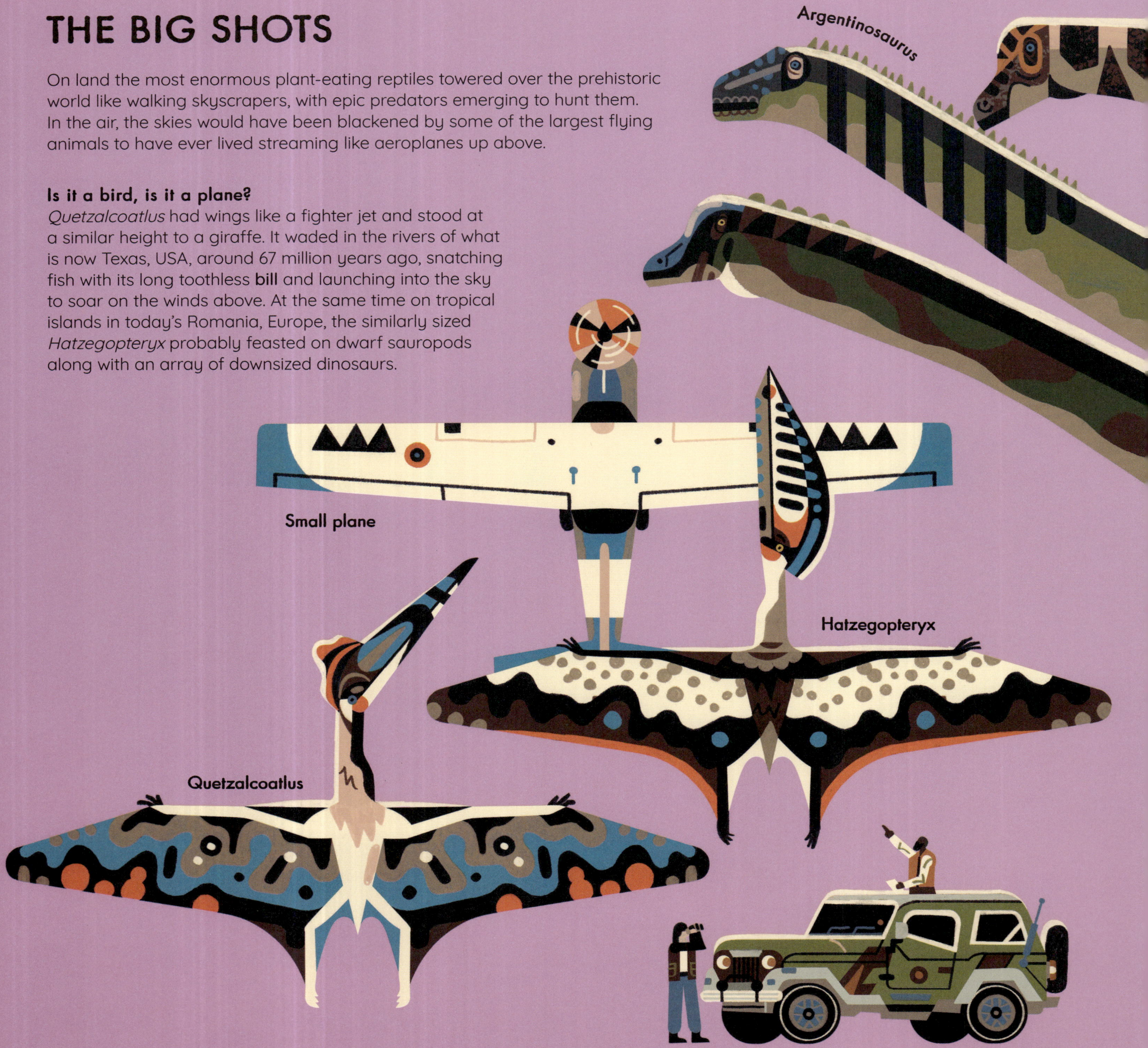

Argentinosaurus

Small plane

Hatzegopteryx

Quetzalcoatlus

Hadrosaurs

Another family of hungry herbivores called the **hadrosaurs** appeared during the Cretaceous Period. They were also known as duck-billed dinosaurs thanks to their flat, wide mouths. Their jaws were packed with a battery of hundreds of growing teeth, each one ready to replace those that wore out. Herds of hadrosaurs spread all over the world, walking on all fours and standing up on their back legs to pick the highest tasty fruits and plants. Some hadrosaurs had toes that joined to form one large nail, similar to horses' hooves. These 'hooves' probably allowed them to run at high speeds.

In the same sites where *Styracosaurus* fossils were found, paleontologists also discovered armoured *Edmontonia* and *Euoplocephalus*, as well as herds of huge hadrosaurs.

Noisy neighbours

Most hadrosaurs had crests. Just like ceratopsians, there were lots of wild varieties. In the past, people wondered if these crests had special functions, from strange snorkels to breathing fire. Today, it's thought that they could push air through these hollow shapes to create bellowing calls to communicate. Each species evolved different crests that made a unique sound, creating a dinosaur symphony across the Cretaceous Period landscape. The crests may have been brightly coloured for attracting mates, too.

Eggs marks the spot

Edmontosaurus nesting sites suggest that hadrosaurs looked after their babies, feeding and caring for them until they matured. Families were packed closely together in nesting groups, laying 30–40 eggs in a spiral shape on the ground. Hatchlings grew remarkably fast, starting at 1 kilogram and reaching 2,000 kilograms within 10 years. That's heavier than a car! Their crests started out as stubs, developing as they aged. Scientists think they may have returned to the same nesting sites over generations, similar to sea turtles.

TYRANTS AND KINGS

Widely known as the most deadly dinosaur of all, *Tyrannosaurus rex* weighed as much as two male hippos at 7,000 kilograms and measured up to 12 metres long, with tiny two-fingered hands and a bulky skull. It tracked down prey with its piercing, orange-sized eyes and keen sense of smell. Its bone-crushing bite was more powerful than any known land animal today, which it used to tear through its victims and swallow 20-kilogram chunks of flesh whole. *Triceratops*, hadrosaurs and ankylosaurs were all on the menu, as well as other carnivores. It may have also scavenged on already dead animals, using its sheer size to scare away other hungry **scavengers**.

A fearsome family

T. rex was part of the tyrannosaur (meaning 'tyrant lizard') family that conquered North America and Asia around 80–66 million years ago. The 9-metre-long *Albertosaurus* and *Gorgosaurus* were similar but less powerful than *T. rex*, possibly hunting in packs across today's North America during the Late Cretaceous Period. *Tarbosaurus* lurked among the rivers and forests rich with sauropods and other tasty herbivores 74–70 million years ago in what is now China and Mongolia.

True titans

Sauropods gave rise to the biggest of all dinosaurs. Titanosaurs lived worldwide during the end of the Cretaceous Period and dwarf today's largest animals. The South American *Argentinosaurus*, *Patagotitan*, *Puertasaurus* and North American *Alamosaurus* were mountainous marvels measuring longer than a blue whale. Many titanosaurs probably lived in awe-inspiring herds.

Popular pilots

One of the most well-known flying reptiles is the horn-headed *Pteranodon*, which lived around 100–90 million years ago across North America mostly, but also in Europe, Asia and South America. It had large eyes for spotting crustaceans, molluscs and fish, and may have stored food in a pouch below its beak like a pelican. Males were much larger than females with a longer bony crest.

Punching up

Prowling the planet for supersized prey were colossal carnivores such as the powerful *Giganotosaurus*, which lived 112–90 million years ago and boasted a thick skull as long as a door. Its slightly smaller relative, *Mapusaurus*, shared the lush habitat with *Argentinosaurus*, possibly hunting this titan in a clash for the ages.

Air superi...

Pterosaurs possessin... neighbour... headwea... possesse... These pte... present-d...

THE END OF THE DINOS

Around 66 million years ago, Earth closely resembled how it looks today. All kinds of dinosaurs roamed the tropical planet, blissfully unaware that something catastrophic was heading their way. Life was about to undergo another huge extinction event – just like the one that gave rise to these mighty reptiles 186 million years earlier.

NORTH AMERICA

EUROPE & ASIA

SOUTH AMERICA

AFRICA

ANTARCTICA

AUSTRALIA

GOING OUT WITH A BANG

The golden age of the great dinosaurs and other magnificent marvels of the prehistoric age abruptly ended 66 million years ago. A vast, 10-kilometre-wide **asteroid** burst through the sky and crashed into the coast of what is now Mexico. It struck Earth with such enormous force that everything in a 965-kilometre-radius was instantly vaporised. In the ensuing chaos, the skies blackened and mountainous waves and hurricane-force winds swept across the planet. Liquid rock, dust and debris flew up into the air, cooling quickly in the atmosphere then raining down as deadly glassy spears.

Acid rain fell and wildfires tore through the forests. Through a combination of luck and resourcefulness, some survivors likely took refuge in burrows or underwater whilst chaos reigned above.

TRAPPED IN TIME

Scientists are able to piece together the dinosaurs' story thanks to fossils. New finds unearthed every year give us a clearer insight into the Mesozoic Era. Some fossils are complete skeletons, but many are bits and pieces, found here and there. From single pelvis bones to pieces of spine or broken teeth, scientists called **palaeontologists** study fossils like prehistoric detectives piecing together a complex puzzle of past lives.

Mary Anning

Myths and monsters

For as long as humans have been on Earth, we have been finding dinosaur fossils. It's possible some of these fossils inspired our earliest tales of mythical beasts, supernatural monsters and biblical giants! Over time, generations of scientists have studied fossils and learned more about them, coming to a better understanding of dinosaurs. Many perfectly complete skeletons now fill our museums.

How is a fossil formed?

When a living thing on land or in the sea dies, it is sometimes buried in bits of rock, soil or mud called sediment. Over time, this turns to solid rock. More layers of earth build up on top, squeezing everything together in layers. Hard bones or shells slowly dissolve away and are replaced by rock too, leaving an imprint – this imprint is called a fossil. The earth and all its layers shift around a lot or wear away, so fossils can become exposed or are found when people dig into the earth.

A needle in a haystack

Fossils are rare. Most living things simply rot away and disappear when they die, especially on land. Some fossils show incredible imprints of skin and flesh or animals locked together in battle. About 99% of fossils come from ocean life, as marine creatures often sink and are quickly buried by sediment on the seafloor. This is why ammonite fossils are found all over the world.

A fall from grace
The next few years were dark and cold. In the oceans, various fish, sharks, shellfish and others escaped the fate of the biggest marine reptiles, who disappeared. On land, rat-sized mammals escaped along with some turtles, crocodiles and small reptiles. What was once a lush planet teeming with extraordinary life turned into a barren world with scattered survivors. Around 80% of life went extinct. The dinosaurs were almost gone.

The survivors

Today's dinosaurs
Technically, dinosaurs are still alive today. While crocodiles share a distant ancestral link with dinosaurs and are close relatives, modern birds are directly descended from theropods. Many birds today started as feathered dinosaurs such as *Archaeopteryx*, retaining the theropods' signature light, hollow bones and air sacs, which were fantastic for flight. A few duck-like dinosaur species survived the extinction by scavenging for scraps. As the Earth recovered and food became more abundant, they thrived and adapted. These eventually evolved into birds, spreading far and wide to fill our skies. This makes the humble chicken the closest living relative to the revered *T. rex*.

The marvellous Mary Anning
The coastline in Dorset, England boasts a rich fossil bed set into the rocks. Mary Anning searched for fossils here since she was a little girl in the 1800s, assembling and studying them to discover new species. Although often uncredited because she was a woman from a modest background, she found the first complete ichthyosaur, many marine reptiles and *Pterodactylus*. People still flock to her stomping grounds, now known as the Jurassic coast, in search of countless fossil curiosities.

Tales through time

Ancient animals, plants, rocks and more are displayed in museums around the world, with over a billion objects on show. Among these are some of the most remarkable fossils, offering a glimpse into how dinosaurs looked and lived.

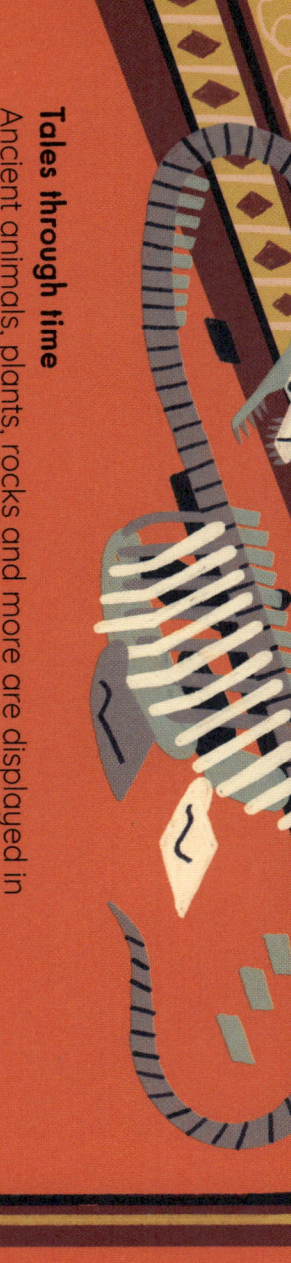

Fossil footprints can tell us about how dinosaurs walked and where they lived. In France, fossilised runways show that pterosaurs landed just like today's birds.

Some fossils show detailed feathers and scientists can even tell what colour or basic patterns certain dinosaurs had.

This 5.5-metre-long *Borealopelta* is preserved with pristine armour, skin and its last meal of ferns. This dinosaur is a Canadian cousin of the armoured *Sauropelta*.

Some dinosaurs' last moments were spent fighting, such as this *Protoceratops* and *Velociraptor* found mid-battle in the Gobi Desert, Asia.

Fossilised eggs and nests tell us how some dinosaurs cared for their young.

Fossilised poo, known as **coprolites**, are key to understanding the diet of some dinosaurs. Millions of years have got rid of any foul smells, thankfully.

Some trees ooze sticky liquid, trapping small creatures, bones, feathers or soft tissue. It hardens into **amber**, which can preserve items for millions of years.

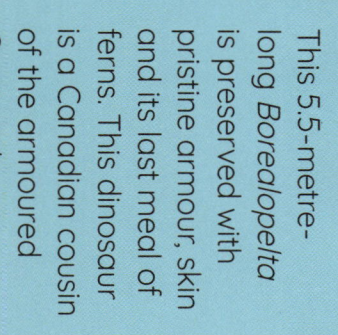

Unsolved mysteries

There are still so many mysteries about dinosaurs. For example, what was the biggest of all time? Which were fluffy or feathered? Which fossils were not pieced together or discovered – were they really? Perhaps one of budding palaeontologists questions in the years to come.

FAMILY TIES

Exactly how, when and why dinosaurs evolved is an ever-changing science, hotly debated among palaeontologists. From the first amphibians to the fluttering birds and cruel crocodiles, this family tree includes many of the dinosaurs covered in this book.

A fish out of water

Our amphibian ancestors evolved from ancient fish around 365 million years ago, slowly swapping fins for fingers and developing strong air-breathing lungs. It is unknown why a handful of fish took to moving on land, but it set in motion an extraordinary evolutionary journey.

Out with the old, in with the new

After the dinosaurs disappeared, mammals took their place and evolved for the next 66 million years (even less time than the Cretaceous Period alone). Eventually, simple, small mammals evolved into humongous ground sloths, bear-sized killer pigs, fluffy mammoths, sabertooth cats, car-sized armoured armadillos and ingenious apes, alongside countless other creatures.

The Bone Wars

people have been studying thousands of years, the word ology' only appeared in English 200 years ago. At this time, logy became a nasty field of d competition at first. A furious 1800s called The Bone Wars 5 years between two wealthy en who despised each other. ed armies of bone-hunters, s, stole or destroyed bones ht to find fossils first. It didn't with both men running out of nd earning bad reputations. this, they found and named us, Allosaurus and Triceratops, among many more.

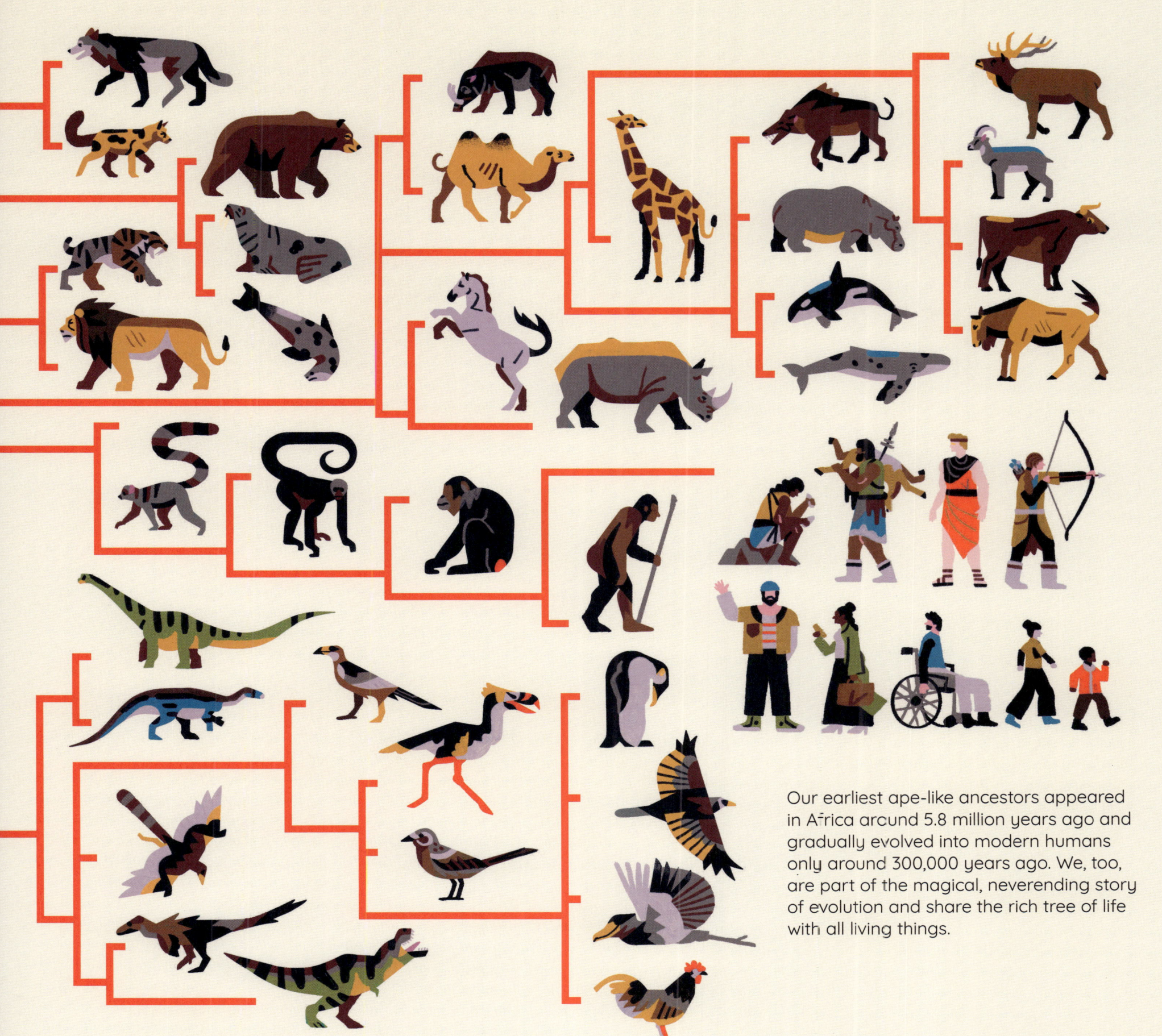

Our earliest ape-like ancestors appeared in Africa around 5.8 million years ago and gradually evolved into modern humans only around 300,000 years ago. We, too, are part of the magical, neverending story of evolution and share the rich tree of life with all living things.

THE DINOSAURS' LEGACY

On the one hand, the age of reptiles was a world of wonder teeming with all kinds of reptiles. Their 186-million-year reign perfectly captures how beautiful and bizarre evolution can be. Dinosaurs have also captured the imagination of adults and children alike – inspiring today's raptor-packed Hollywood blockbuster movies and sparking the fossil-finding fever that gripped the world 200 years ago.

Atlas bear

Passenger pigeon

Dodo

Gone, but not forgotten
On the other hand, the dinosaurs' story is a stark warning. When *T. rex* awoke 66 million years ago on an abundant planet where it reigned supreme, it didn't know the terror about to befall it. We do. Unfortunately, many animals today are at risk of repeating the dinosaurs' fate. Climate change, unsustainable farming, rising sea levels, overhunting and habitat loss have pushed thousands of animals to the brink of extinction.

Thylacine

The atlas bear, passenger pigeon, dodo and thylacine, are among the many creatures lost in the past few hundred years.

Habitat hero

Rising global temperatures are harming people and animals. There are lots of things you can do, big or small, to help our planet and the creatures around us. These include:

Avoid waste, and recycle | **Plant trees or local plants** | **Walk, cycle or use public transport**

Use less water | **Eat a more plant-based diet** | **Buy from sustainable companies**

Spread the word

Talk about the planet, peacefully protest or join in with the noble people and organisations around the world who work tirelessly to protect our precious wildlife today.

GLOSSARY

Amber – Hardened sap from ancient trees that has become solid over a long time. Usually yellow or orange in colour.

Ambusher – Something that attacks by surprising its victim.

Ankylosaur – A group of armoured, plant-eating dinosaurs that lived during the Cretaceous Period.

Asteroid – A rock that orbits the sun.

Bill – The mouthpart of a bird, also known as a beak.

Carnivore – An animal that feeds on other animals. *Allosaurus*, *Giganotosaurus* and *Tyrannosaurus* are carnivores.

Coprolite – Fossilised poop.

Crest – A tuft of feathers or fur, or a ridge, that some birds or reptiles have atop their heads.

Cretaceous Period – The last period of the Mesozoic Era, from 145 to 66 million years ago.

Dinosaur – A group of reptiles of all shapes and sizes that dominated the land during the Mesozoic Era.

Evolution – A process by which living things change over time.

Extinction – When a group of animals or plants dies out completely.

Food chain – A process where living things rely on each other for food.

Fossil – The remains of a living thing, such as a plant or animal, usually preserved in rock.

Frill – A tuft of feathers, hair or bony growths on the neck of an animal.

Genus (plural: genera) – Scientists use different names to organise different kinds of animals. Genus is a group of varied but closely related animals. For example, horses and zebras are part of the genus *Equus*.

Hadrosaur – A group of large, four-legged, plant-eating dinosaurs from the Cretaceous Period. They are also known as duck-billed dinosaurs thanks to their flat, beak-like mouths.

Herbivore – An animal that only eats plants. *Triceratops* and *Stegosaurus* were herbivores.

Herd – A large group of animals that live or travel together.

Insular dwarfism – A process where a small group of large animals that are limited to one remote area (usually an island) become smaller over time.

Jurassic Period – The period from 201 to 145 million years ago.

Keratin – An important substance that helps form scales, hair, feathers, nails, hooves and horns.

Mesozoic Era – The entire timespan of the dinosaurs' reign, from 252 to 66 million years ago.

Mollusc – A wide-ranging, soft-bodied group of living organisms. Oysters, snails, ammonites and octopuses are molluscs.

Mosasaur – A family of large, predatory marine reptiles that swam in the oceans during the Cretaceous Period.

Ornithischian – A group of mostly plant-eating dinosaurs known for their bird-like pelvis bones. *Stegosaurus* and *Triceratops* are ornithischians.

Osteoderm – A hard, bony structure that can form scales, plate armour or shells.

Pachycephalosaur – A group of plant-eating dinosaurs from the Cretaceous Period, known for their bony heads.

Palaeontologist – A scientist who studies fossils and ancient life.

Pangaea – The name of the giant landmass on Earth that existed from around 300–200 million years ago. It subsequently split up into today's continents.

Panthalassa – The gigantic ocean that surrounded Pangaea.

Predator – An animal or organism that kills other animals.

Prey – An animal or organism that is killed by other animals.

Reptile – A cold-blooded animal, usually scaly. Crocodiles, snakes and dinosaurs are reptiles.

Sail – A large, flat extension of an animal's spine that sticks out. *Spinosaurus* has a sail on its back.

Saurischian – A group of dinosaurs known for their lizard-like pelvis bones. *Diplodocus* and *Allosaurus* are saurischians.

Sauropod – A group of plant-eating dinosaurs known for their long necks and massive sizes.

Scavenger – Creatures that feed on the remains of dead animals or leftover plants.

Species (see 'genus') – A group of closely related living things within a genus that can all breed together to produce fertile offspring.

Theropod – A group of carnivorous saurischian dinosaurs that typically walk on two legs and have short front arms. The mighty *Tyrannosaurus* is a theropod.

Triassic Period – The period from 252 to 201 million years ago.

PRONUNCIATION GUIDE

Abelosaurid (ah-beel-oh-sore-id)
Achelousaurus (ah-kel-oo-sore-us)
Aetosaur (ay-ee-toh-sor)
Alamosaurus (ah-la-mow-sore-us)
Allosaurus (al-oh-sore-us)
Amargasaurus (a-marg-oh-scre-us)
Ankylosaurus (an-kye-low-sore-us)
Anurognathus (an-uh-rog-nath-us)
Apatosaurus (ah-pat-oh-sore-us)
Archaeopteryx (ark-ee-opt-er-ix)
Archosaur (ark-oh-sore)
Avimimus (ah-vee-meem-us)
Balaenognathus (bay-lee-nuh-nah-thus)
Caelestiventus (kay-el-es-te-ven-tus)
Camarasaurus (kam-ar-uh-sore-us)
Carcharodontosaurus (kar-kar-oh-don-toe-sore-us)
Carnotaurus (kar-noh-tore-us)
Centrosaurus (cen-troh-sore-us)
Ceratopsian (serr-uh-top-see-un)
Ceratosaurus (keh-rat-oh-sore-us)
Chasmosaurus (kaz-mo-sore-us)
Chungkingosaurus (chung-chirg-oh-sore-us)
Coelophysis (seel-oh-fie-sis)
Compsognathus (komp-sog-nath-us)
Concavenator (kon-ka-ven-at-or)
Corythosaurus (koh-rith-oh-sore-us)
Cretaceous (kruh-tay-shuhs)
Cretoxyrhina (kree-tox-ee-rye-nah)
Cryolophosaurus (cry-o-loaf-oh-sore-us)
Cynognathus (sih-nog-nay-thus)
Deinonychus (dye-non-ick-us)
Dilophosaurus (die-loaf-oh-sore-us)
Dimorphodon (dee-morf-oh-don)
Edmontonia (ed-mon-tone-ee-ah)
Einiosaurus (ie-nee-oh-sore-us)
Eoraptor (ee-oh-rap-tor)
Eudimorphodon (you-di-mor-fo-don)

Euoplocephalus (you-oh-plo-kef-ah-luss)
Europasaurus (yoo-roh-pah-sore-us)
Gallimimus (galley-mime-us)
Gargoyleosaurus (gahr-goy-lee-oh-sore-us)
Gastonia (gas-toh-nee-ah)
Giganotosaurus (gig-an-oh-toe-sore-us)
Gigantoraptor (ji-gan-to-rap-tor)
Gorgosaurus (gor-goh-sore-us)
Gryposaurus (grip-oh-sore-us)
Hatzegopteryx (hat-suh-gop-ter-ix)
Herrerasaurus (herr-ray-rah-sore-us)
Heterodontosaurus (het-er-oh-dont-oh-sore-us)
Ichthyosaur (ik-thee-uh-sore)
Incisivosaurus (in-sih-sih-voh-sore-us)
Kosmoceratops (kos-moh-serra-tops)
Kronosaurus (kron-oh-sore-us)
Liliensternus (lil-ee-en-shtern-us)
Lystrosaurus (lis-tro-sore-us)
Maiasaura (my-ah-sore-ah)
Mapusaurus (mah-puh-sore-us)
Mesozoic (meh-suh-zow-ic)
Microraptor (mike-roe-rap-tor)
Nasutoceratops (nah-soo-toh-serra-tops)
Olorotitan (oh-lore-oh-tye-tan)
Ornithischia (or-nith-is-kee-uh)
Oviraptor (oh-vee-rap-tuh)
Oxalaia (ox-ah-lie-ah)
Pachycephalosaurus (pack-ee-seff-ah-lo-sore-us)
Palaeontology (pay-lee-uhn-to-luh-jee)
Pangaea (pan-jee-uh)
Parasaurolophus (pa-ra-saw-rol-off-us)
Patagotitan (pat-ah-go-tie-tan)
Peteinosaurus (peh-ty-no-sore-us)
Plateosaurus (plat-ee-oh-sore-us)
Plesiosaur (plee-zee-uh-sore)
Pliosaur (pligh-oh-sore)

Prosaurolophus (proh-sore-oh-lof-us)
Protoceratops (pro-toe-ker-ah-tops)
Psittacosaurus (sit-ak-oh-sore-us)
Pteranodon (ter-an-uh-don)
Pterodactylus (ter-uh-dak-tuhl)
Puertasaurus (pwer-ta-sore-us)
Quetzalcoatlus (ket-suhl-koh-at-luss)
Rajasaurus (rah-juh-sore-us)
Rhyncosaurs (ring-koh-sore)
Sarahsaurus (sah-rah-sore-us)
Saurischia (sore-is-kee-uh)
Sauropelta (sore-oh-pelt-ah)
Sauropod (sore-ruh-pod)
Saurosuchus (sore-oh-soo-kuss)
Scutellosaurus (skoo-tell-oh-sore-us)
Shantungosaurus (shan-tun-go-sore-us)
Shunosaurus (shoon-oh-sore-us)
Sinoceratops (sigh-no-serra-tops)
Spinosaurus (spine-oh-sore-us)
Squalicorax (skwahl-i-cor-ax)
Stegosaurus (steg-oh-sore-us)
Styracosaurus (sty-rak-oh-sore-us)
Tanystropheus (tan-ee-stro-fee-us)
Tarbosaurus (tar-bow-sore-us)
Therizinosaurus (ther-ih-zine-oh-sore-us)
Thylacine (thy-uh-seen)
Torvosaurus (tor-voh-sore-us)
Triceratops (tri-serra-tops)
Tropeognathus (tro-pee-og-na-thus)
Tsintaosaurus (ching-dow-sore-us)
Tuojiangosaurus (too-yang-oh-sore-us)
Tupandactylus (too-pan-dak-tuhl-us)
Tupuxuara (too-poo-hwar-a)
Tyrannosaur (tie-ran-oh-sore-us)
Velociraptor (vel-oss-ee-rap-tor)
Xiphactinus (zih-fact-in-us)
Yangchuanosaurus (yang-choo-ahn-oh-sore-us)